# How to be Brilliant at

# ALGEBRA

Beryl Webber
Terry Barnes

 Brilliant Publications

We hope you and your class enjoy using this book. Other books in the series include:

*Science titles*
How to be Brilliant at Recording in Science     978 1 897675 10 6
How to be Brilliant at Science Investigations     978 1 897675 11 3
How to be Brilliant at Materials     978 1 897675 12 0
How to be Brilliant at Electricity, Light and Sound     978 1 897675 13 7
How to be Brilliant at Living Things     978 1 897675 66 3

*English titles*
How to be Brilliant at Writing Stories     978 1 897675 00 7
How to be Brilliant at Writing Poetry     978 1 897675 01 4
How to be Brilliant at Grammar     978 1 897675 02 1
How to be Brilliant at Making Books     978 1 897675 03 8
How to be Brilliant at Spelling     978 1 897675 08 3
How to be Brilliant at Reading     978 1 897675 09 0
How to be Brilliant at Word Puzzles     978 1 897675 88 5

*Maths titles*
How to be Brilliant at Using a Calculator     978 1 897675 04 5
How to be Brilliant at Numbers     978 1 897675 06 9
How to be Brilliant at Shape and Space     978 1 897675 07 6
How to be Brilliant at Mental Arithmetic     978 1 897675 21 2

*History and Geography titles*
How to be Brilliant at Recording in History     978 1 897675 22 9
How to be Brilliant at Recording in Geography     978 1 897675 31 1

*Christmas title*
How to be Brilliant at Christmas Time     978 1 897675 63 2

Published by Brilliant Publications,
Unit 10, Sparrow Hall Farm,
Edlesborough,
Dunstable,
Bedfordshire,
LU6 2ES

Sales and stock enquiries:
Tel:    01202 712910
Fax:    0845 1309300
e-mail:brilliant@bebc.co.uk
www.brilliantpublications.co.uk

General information enquiries:
Tel:    01525 222292
The name 'Brilliant Publications' and its logo are
registered trademarks.

Written by Beryl Webber and Terry Barnes
Illustrated by Kate Ford
Cover photograph by Martyn Chillmaid

Printed in the UK

# Contents

# Introduction

*How to be Brilliant at Algebra* contains 42 photocopiable sheets for use with 7–11 year olds. The ideas are written in line with the National Curriculum programmes of study and attainment targets. They can be used whenever the need arises for particular activities to support and supplement whatever core mathematics programme you follow. The activities provide learning experiences that can be tailored to meet individual children's needs.

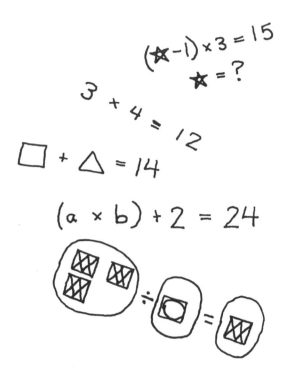

The activities are addressed directly to the children. They are self-contained and many children will be able to work with very little additional support from you. You may have some children, however, who have the necessary mathematical concepts and skills, but require your help in reading the sheets.

The children will need pencils and should be encouraged to use the sheets for all their working. They may require a calculator for some of the activities. Other activities require additional basic classroom mathematics resources, such as counters, cubes and dice.

A few of the activities require the use of additional resource sheets and these are given at the back of the book. Where this is the case it has been indicated by a small box, with the page number in it, in the top right corner of the sheet, eg 47 .

*How to be Brilliant at Algebra* relates directly to the programmes of study for algebra and using and applying mathematics. The page opposite gives further details and on the contents page the activities are coded according to attainment target(s) and level(s).

Page 46 provides a self-assessment sheet so that children can keep a record of their own progress.

# Links to the National Curriculum

**The activities in this book allow children to have opportunities to:**

- use and apply mathematics in practical tasks, in real-life problems and within mathematics itself;

- take increasing responsibility for organizing and extending tasks;

- devise and refine their own ways of recording;

- ask questions;

- develop flexible and effective methods of computation and recording, and use them with understanding;

- use calculators for exploring number structure;

- consider a range of patterns.

In particular, the activities relate to the following sections of the Key Stage 2 programme of study:

## Using and Applying Mathematics

**2.** **Making and monitoring decisions to solve problems**
    **a**     select and use the appropriate mathematics and materials;

    **b**     try different mathematical approaches;

    **c**     develop their own mathematical strategies and look for ways to overcome difficulties;

    **d**     check their results and consider whether they are reasonable.

**3.** **Developing mathematical language and forms of communication**
    **a**     understand and use the language of:
        - number
        - relationships;

    **b**     use diagrams, graphs and simple algebraic symbols;

    **c**     present information and results clearly.

**4.** **Developing mathematical reasoning**
    **a**     understand and investigate general statements;

    **b**     search for pattern in their results;

    **c**     make general statements of their own, based on evidence they have produced;

    **d**     explain their reasoning.

## Numbers

**3.** **Understanding relationships between numbers and developing methods of computation**
    **a**     explore number sequences, explain patterns and use simple relationships; interpret, generalize and use simple mappings;

    **b**     recognize number relationships;

    **c**     use some properties of numbers, including multiples, factors, squares and primes;

    **d**     develop a variety of mental methods of computation and explain patterns used in all four operations;

**f** understand and use the relationship between the four operations, including inverses;

**g** extend methods of computation to include all four operations with decimals, using a calculator where appropriate.

## 4. Solving numerical problems

**a** develop their use of the four operations to solve problems;

**c** check results by different methods.

## Shape, Space and Measures

## 2. Understanding and using properties of shape

**a** visualize and describe shapes and movements, developing precision in using related geometrical language.

# Patterns

Many patterns exist in our world. Collect some examples of fabric, wallpaper and crockery. Look carefully at the patterns and describe what you see.

> **Tip**: These words might help you. Look them up in a dictionary if you are not sure what they mean:
>
> | | |
> |---|---|
> | patterns | transformation |
> | motif | translation |
> | repetition | reflection |
> | rotation | |

Now investigate patterns in the natural world.

You might like to look at:

fir-cones
shells
fruit

sunflower heads
butterflies
leaves

pineapples
petals
vegetables

Draw some of the patterns you've found here:

---

**EXTRA!**
Design wrapping paper and make a matching tag. First choose a pattern to use.
You may find it easier to use squared paper. Print your pattern using thick paint.

---

How to be Brilliant at Algebra

# Sweets

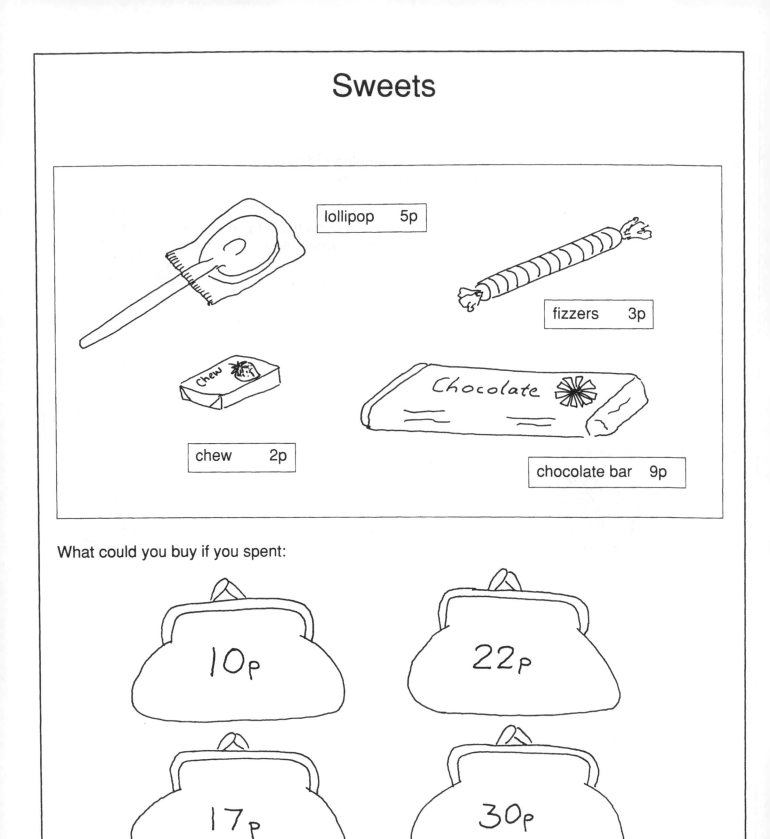

lollipop 5p

fizzers 3p

chew 2p

chocolate bar 9p

What could you buy if you spent:

10p

22p

17p

30p

How to be Brilliant at Algebra

# Inverse operations

Add, subtract, multiply and divide are sometimes called **operations**. An **inverse operation** is an operation that will take you back to the number you started with.

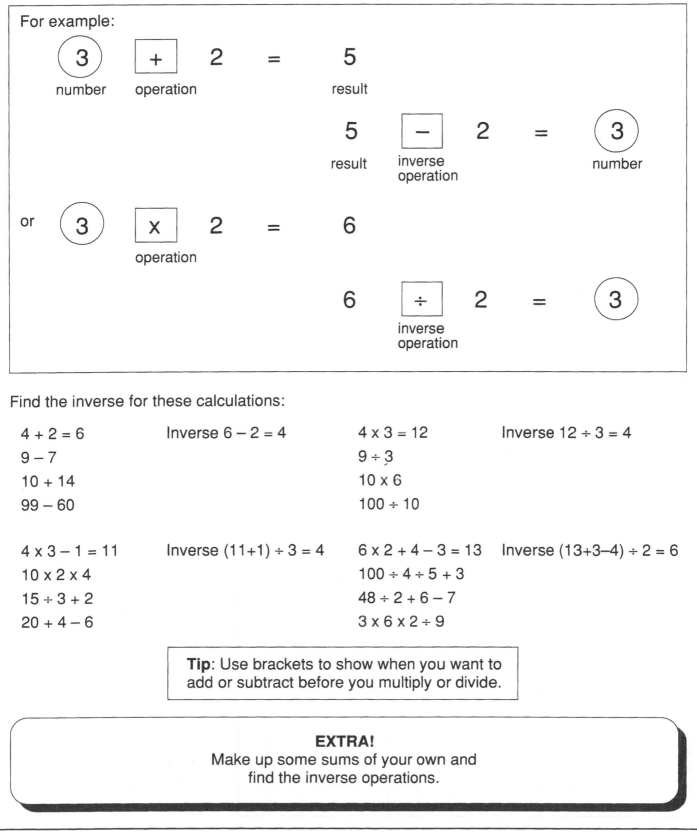

For example:

3    +    2    =    5

number    operation        result

5    −    2    =    3

result    inverse operation    number

or    3    x    2    =    6

operation

6    ÷    2    =    3

inverse operation

Find the inverse for these calculations:

4 + 2 = 6      Inverse 6 − 2 = 4      4 x 3 = 12      Inverse 12 ÷ 3 = 4

9 − 7                                       9 ÷ 3

10 + 14                                  10 x 6

99 − 60                                 100 ÷ 10

4 x 3 − 1 = 11      Inverse (11+1) ÷ 3 = 4      6 x 2 + 4 − 3 = 13      Inverse (13+3−4) ÷ 2 = 6

10 x 2 x 4                                  100 ÷ 4 ÷ 5 + 3

15 ÷ 3 + 2                                48 ÷ 2 + 6 − 7

20 + 4 − 6                                3 x 6 x 2 ÷ 9

> **Tip**: Use brackets to show when you want to add or subtract before you multiply or divide.

**EXTRA!**
Make up some sums of your own and
find the inverse operations.

# Cubes

Choose three colours of interlocking cubes. Investigate how many different groups of three cubes you can make using all three colours each time.

Record your groups here:

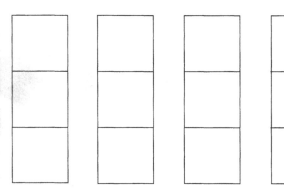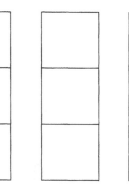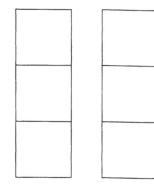

Try to order them in a logical way.

Try again. This time you can use each colour more than once in a group. Work systematically. Record each new group here as you work.

**Tip**: You could colour in the squares or you could use a letter code for each colour, for example:

**R** for red
**Y** for yellow
**G** for green

# Odds and evens

*If you use dice with more than six sides the activity will be even more interesting!*

You will need five dice.

Throw two dice. Total the dots. Record your throws and the total in the table. Repeat several times.

| Die 1 | Die 2 | Total |
|-------|-------|-------|
|       |       |       |
|       |       |       |
|       |       |       |
|       |       |       |
|       |       |       |
|       |       |       |
|       |       |       |
|       |       |       |

Circle all even numbers in red and all odd numbers in blue. Look for a pattern.

Repeat using three dice.

| Die 1 | Die 2 | Die 3 | Total |
|-------|-------|-------|-------|
|       |       |       |       |
|       |       |       |       |
|       |       |       |       |
|       |       |       |       |
|       |       |       |       |
|       |       |       |       |
|       |       |       |       |
|       |       |       |       |

Look for a pattern. What do you notice?

Try again with four dice and five dice. What do you notice?

**Tip**: Make a table showing the results. Use **O** for an odd number and **E** for even numbers. For example:

**O** + **O** =
**O** + **E** =
**E** + **O** =
**E** + **E** =

### EXTRA!
Predict what would happen with six and seven dice. Try it and see.

# Two dozen, 1

Play this game with a friend. The goal is to reach two dozen (24).

You will need some counters of two colours and one copy of this sheet between you. Each choose a colour. Take it in turns to place a counter on the grid. Only one counter is allowed on each number. The first player to have counters on numbers which total 24 is the winner.

| | | | | |
|---|---|---|---|---|
| 1 | 2 | 3 | 4 | 5 |
| 6 | 7 | 8 | 9 | 10 |
| 11 | 12 | 13 | 14 | 15 |

Play the game several times and record the winning combinations here.

Investigate:

◆ how many ways you can win using only 2 counters

◆ how many ways you can win using 3 counters

◆ how many ways you can win using 4 counters

---

**EXTRA!**
Investigate the patterns made by the winning combinations.

---

How to be Brilliant at Algebra

# Two dozen, 2

Record your results on a separate sheet of paper. You will need plenty.

Investigate ways of making 24 using 2 numbers.

☐ + ☐ = (24)

Investigate different ways of completing this sum. Hold the 8 constant.

☐ + (8) + ☐ = (24)

For each different number you hold constant, how many ways can you total 24?
Is there a pattern?

☐ + ○ + ☐ = (24)

Investigate adding 4 numbers to total 24.

☐ + ☐ + ☐ + ☐ = (24)

Try holding two numbers constant. Is there a pattern?

☐ + ○ + ○ + ☐ = (24)

---

**EXTRA!**
Choose another number to investigate in this way.

---

How to be Brilliant at Algebra

# Find the missing numbers

Can you find the missing numbers? Write the correct sum underneath the question.

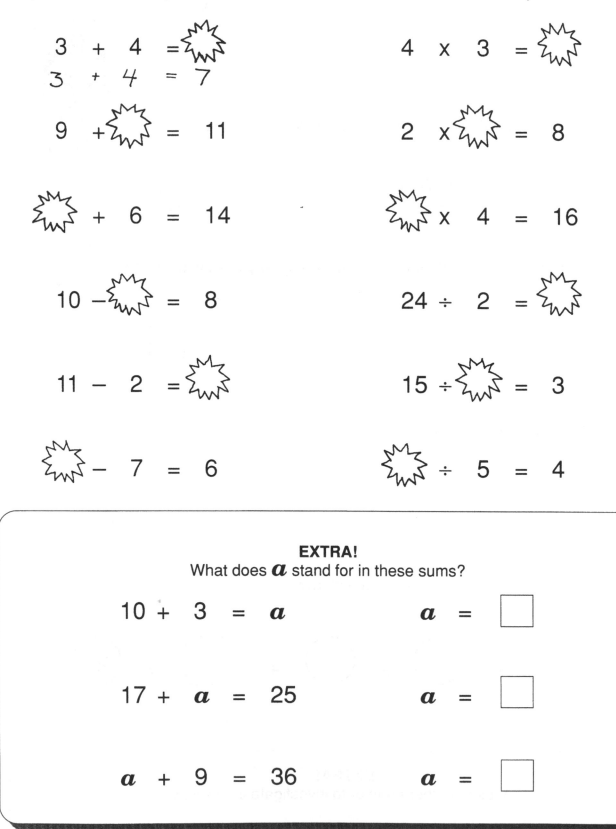

3 + 4 = ✸

3 + 4 = 7

4 x 3 = ✸

9 + ✸ = 11

2 x ✸ = 8

✸ + 6 = 14

✸ x 4 = 16

10 − ✸ = 8

24 ÷ 2 = ✸

11 − 2 = ✸

15 ÷ ✸ = 3

✸ − 7 = 6

✸ ÷ 5 = 4

**EXTRA!**
What does $a$ stand for in these sums?

10 + 3 = $a$        $a$ = ☐

17 + $a$ = 25        $a$ = ☐

$a$ + 9 = 36        $a$ = ☐

# Variables

Investigate the numbers that could make these sums correct. Record as many different solutions as you can find for each sum.

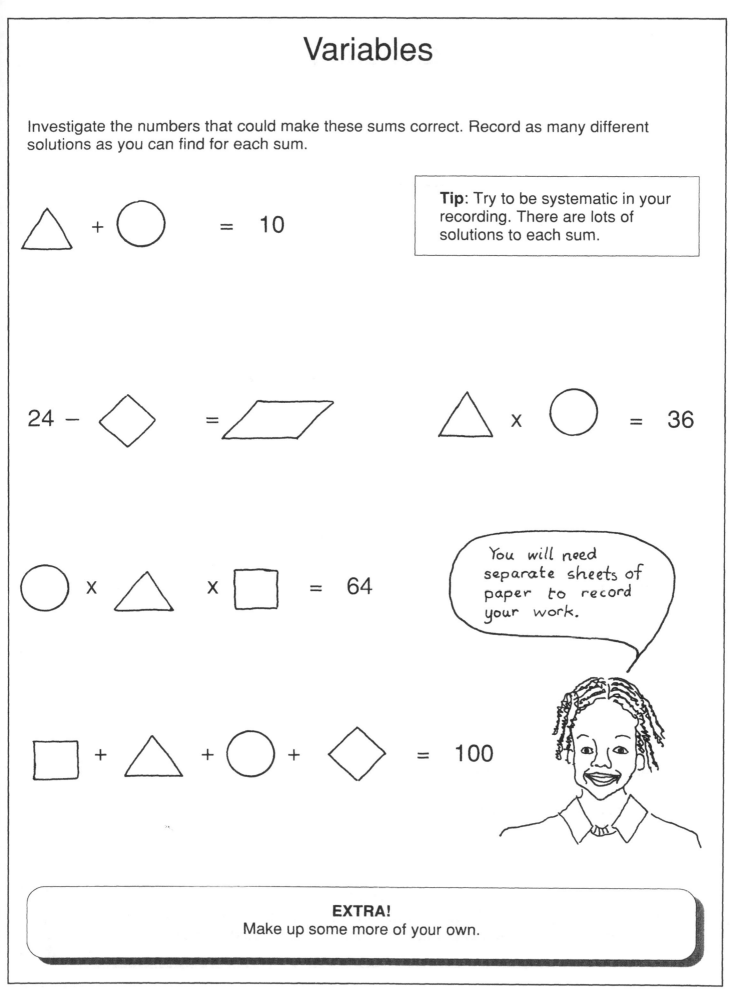

**Tip**: Try to be systematic in your recording. There are lots of solutions to each sum.

$\triangle + \bigcirc = 10$

$24 - \diamondsuit = \diagup\!\!\!\!\diagdown$

$\triangle \times \bigcirc = 36$

$\bigcirc \times \triangle \times \square = 64$

You will need separate sheets of paper to record your work.

$\square + \triangle + \bigcirc + \diamondsuit = 100$

**EXTRA!**
Make up some more of your own.

How to be Brilliant at Algebra

Calendars are a rich source of number patterns. The calendar below shows a month of 31 days.

| S | M | T | W | T | F | S |
|---|---|---|---|---|---|---|
|   |   |   | 1 | 2 | 3 | 4 |
| 5 | 6 | 7 | 8 | 9 | 10 | 11 |
| 12 | 13 | 14 | 15 | 16 | 17 | 18 |
| 19 | 20 | 21 | 22 | 23 | 24 | 25 |
| 26 | 27 | 28 | 29 | 30 | 31 |   |

Use counters to cover the following numbers and see what patterns you get. Record the patterns on the calendar resource sheet (page 47).

- numbers in the 2 times table

- numbers in the 3 times table

- numbers in the 4 times table

- numbers in the 5 times table

- numbers in the 6 times table

- numbers in the 7 times table

**EXTRA!**
Investigate the patterns with another month. You could use the second month on the resource sheet or perhaps the month of your birthday this year.

How to be Brilliant at Algebra

# Calendars, 2

There are interesting number patterns in calendars. The calendar below shows
a month of 31 days.

| S | M | T | W | T | F | S |
|---|---|---|---|---|---|---|
|   |   |   |   | 1 | 2 | 3 |
| 4 | 5 | 6 | 7 | 8 | 9 | 10 |
| 11 | 12 | 13 | 14 | 15 | 16 | 17 |
| 18 | 19 | 20 | 21 | 22 | 23 | 24 |
| 25 | 26 | 27 | 28 | 29 | 30 | 31 |

Use counters to cover the following number pattern. Describe the pattern you see.

■ 1,  8,  15,  22,  29

Complete these patterns. What is the difference between each number in the series?

■ 3,  10,  17,  ☐ ,  ☐        The difference is ☐ .

■ 2,  8,  14,  20,  ☐        The difference is ☐ .

■ 1,  7,  13,  ☐ ,  ☐        The difference is ☐ .

■ 1,  9,  17,  25        The difference is ☐ .

■ 5,  13,  ☐ ,  ☐        The difference is ☐ .

---

**EXTRA!**
Make up two more patterns, one with a difference of 5 and one with a difference of 9.
You can use this calendar or you may wish to choose a different month.

---

How to be Brilliant at Algebra

# Calendars, 3

You will need a calendar for this activity. Use the Calendar resource sheet (page 47) or a calendar of your own.

1  Investigate patterns of four squares.

For example:

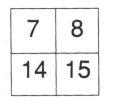

|    |    |
|----|----|
| 7  | 8  |
| 14 | 15 |

What do you notice about the totals of the diagonals, rows and columns?

Does this work for any 2 x 2 square on a calendar?

2  Investigate patterns of nine squares.

For example:

| 15 | 16 | 17 |
|----|----|----|
| 22 | 23 | 24 |
| 29 | 30 | 31 |

Investigate:

- • the totals of opposite corners
- • the totals of the diagonals
- • the totals of the middle row and column
- • the relationship between the total of the diagonals and the middle number

3  Investigate patterns of sixteen squares.

For example:

| 6  | 7  | 8  | 9  |
|----|----|----|----|
| 13 | 14 | 15 | 16 |
| 20 | 21 | 22 | 23 |
| 27 | 28 | 29 | 30 |

Investigate:

- • the totals of the diagonals
- • opposite corners of the 4 x 4 square and the middle 2 x 2 square

---

**EXTRA!**
Choose some other squares from a calendar and investigate some ideas of your own.

---

How to be Brilliant at Algebra

# Number patterns, 1

Look at this number pattern. Each term is two more than the one before.
We say the pattern has a 'difference of 2'.

2        4        6·        8

Investigate the number patterns below. Find the difference in each case and use that
information to calculate the tenth term in the series.

Tenth term

| | | | | | | Tenth term |
|---|---|---|---|---|---|---|
| 1 | 3 | 5 | 7 | 9 | _____ | ☐ |
| 20 | 18 | 16 | 14 | 12 | _____ | ☐ |
| 4 | 7 | 10 | 13 | 16 | _____ | ☐ |
| 5 | 9 | 13 | 17 | 21 | _____ | ☐ |
| 24 | 20 | 16 | 12 | 8 | _____ | ☐ |
| 1 | 11 | 21 | 31 | 41 | _____ | ☐ |
| -3 | 0 | 3 | 6 | 9 | _____ | ☐ |
| 10 | 19 | 28 | 37 | 46 | _____ | ☐ |

**Tip**: Once you know the difference you could use your calculator
to set up a constant function to find the tenth term. For example:

| + | = |          | – | = |
| + | + | = |    | – | – | = |

## EXTRA!
Use your calculator to help you create some number patterns of your own.
Ask a friend to find the tenth term.

How to be Brilliant at Algebra

# Number patterns, 2

In this number pattern each term is twice as much as the one before. It is a 'multiply by 2' pattern.

2        4        8        16

Investigate the patterns below. Find the number each term is multiplied or divided by and use that information to calculate the tenth term in the series.

**Tip**: Once you know the amount of the increase or decrease, you can use a calculator to set up a constant multiplication or division function to find the tenth term.

**Tenth term**

| | | | | | |
|---|---|---|---|---|---|
| 3 | 6 | 12 | 24 | _____ | |
| 3 | 9 | 27 | 81 | _____ | |
| 4 | 12 | 36 | 108 | _____ | |
| 128 | 64 | 32 | 16 | _____ | |
| 1 | 10 | 100 | 1000 | _____ | |

You can combine multiplying or dividing with adding or subtracting, to get different number patterns. For example:

3        5        9        17

is created by 'double the number and take away 1'.

What is the rule for these patterns?

4        9        19        39

10        25        70        205

---

**EXTRA!**
Make up some similar patterns of your own.
Try them on a friend.

# What's the rule?

Work with a friend. You will need two dice and a calculator.

Secretly choose a rule to combine two numbers. For example, you could have:

'add the two numbers together and take away 1'

or

'multiply the first number by 3 and add the second number'

Write the rule down secretly.

Your friend throws the two dice and records the numbers in the grid. You perform your rule using a calculator and write down the answer. Your friend throws the dice again.
Keep going until he can work out what your rule is.

| 1st number (*a*) | 2nd number (*b*) | Answer |
|---|---|---|
|  |  |  |
|  |  |  |
|  |  |  |
|  |  |  |
|  |  |  |
|  |  |  |

Swap roles and play the game again.

**Tip**: If your friend cannot guess the rule, give him a clue.

For example, if the rule is 'multiply the first number by 3 and add the second number' and the dice throws are 3 and 6, tell your friend that the half way stage is 9 (3 x 3) and the answer is 15.

**EXTRA!**
You can write your rules using '*a*' for the first number and '*b*' for the second number.
For example 'add the two numbers together and take away 1'
becomes '*a* + *b* − 1'.

# Totals

Work with a friend. You will need counters of two different colours and a calculator. Each choose a colour.

Look at the grid below. Secretly choose two numbers that are next to each other:

* vertically      ↕  or
* horizontally  ↔  or
* diagonally     ⤢⤡

Add them, using a calculator, and tell your friend the answer. Your friend should try to guess what the numbers are. If she is right she should cover the numbers with her counters and it is then her turn. If she is wrong, you cover the numbers with your counters. It is still your go, so choose two more numbers. Continue until you have covered as much of the grid as possible.

| 1 | 2 | 3 | 4 | 5 | 6 |
|---|---|---|---|---|---|
| 7 | 8 | 9 | 10 | 11 | 12 |
| 13 | 14 | 15 | 16 | 17 | 18 |
| 19 | 20 | 21 | 22 | 23 | 24 |
| 25 | 26 | 27 | 28 | 29 | 30 |
| 31 | 32 | 33 | 34 | 35 | 36 |

The winner is the person with the most numbers covered on the grid.

# Nines, 1

Look at the patterns in the 9 times table. Continue the table up to 10 x 9. Use a calculator to help you.

| | | | | |
|---|---|---|---|---|
| 1 | x | 9 | = | **9** |
| 2 | x | 9 | = | **1 8** |
| 3 | x | 9 | = | **2 7** |
| 4 | x | 9 | = | **3 6** |
| 5 | x | 9 | = | **4 5** |
| 6 | | | = | **5 4** |
| 7 | | | = | **6 3** |
| 8 | | | | |
| 9 | | | | |
| 10 | | | | |

**Tip**: You can set up a constant function on a calculator to help you. Press either:

| 9 | + | = | | or |

| 9 | + | + | = |

Look at the pattern of units in each answer. What do you notice? Now look at the tens digits. What do you notice?

Record the answers on a 100 square. What do you notice? Predict the answer to 11 x 9 and use the calculator to check. Continue the pattern to 20 x 9. Predict 25 x 9 and check.

| | |
|---|---|
| 11 | 16 |
| 12 | 17 |
| 13 | 18 |
| 14 | 19 |
| 15 | 20 |

---

### EXTRA!
A quick way to calculate the 9 times table is to use your hands!
Hold up both hands. Suppose you want to work out 3 x 9. Bend down your third finger from the left. Count how many figures are still standing on the left of the bent finger. This is the tens number. Now count the fingers to the right of the bent finger. This is the units number.

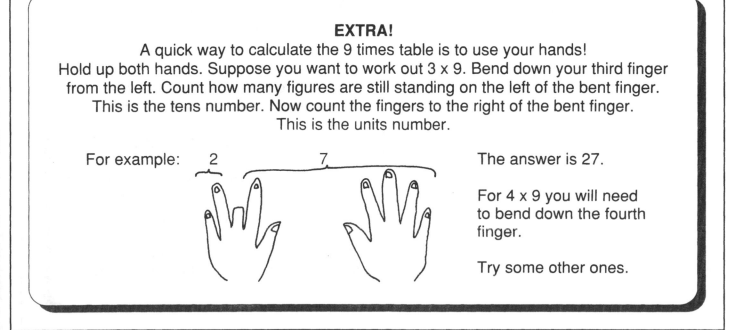

For example:  2  7  The answer is 27.

For 4 x 9 you will need to bend down the fourth finger.

Try some other ones.

How to be Brilliant at Algebra

# Nines, 2

Nine is a magical number. Add up the digits of the answers in the 9 times table.
The first three have been done for you. Look for the magical nine:

| | | | | | | | | | |
|---|---|---|---|---|---|---|---|---|---|
| 1 x 9 | = | 9 | 9 | = | 9 | 6 x 9 = | 54 | | |
| 2 x 9 | = | 18 | 1 + 8 | = | 9 | 7 x 9 = | 63 | | |
| 3 x 9 | = | 27 | 2 + 7 | = | 9 | 8 x 9 = | 72 | | |
| 4 x 9 | = | 36 | | | | 9 x 9 = | 81 | | |
| 5 x 9 | = | 45 | | | | 10 x 9 = | 90 | | |

The total you get when you add all the digits in a number is called the digital root.
What happens with larger numbers in the 9 times table?

| | | | | | |
|---|---|---|---|---|---|
| 11 x 9 = | 99 | | 16 x 9 = | 144 | |
| 12 x 9 = | 108 | | 17 x 9 = | 153 | |
| 13 x 9 = | 117 | | 18 x 9 = | 162 | |
| 14 x 9 = | 126 | | 19 x 9 = | 171 | |
| 15 x 9 = | 135 | | 20 x 9 = | 180 | |

**Tip**: If you get a two digit number, try one more time.

$$9 + 9 = 18$$
$$1 + 8 = \mathbf{9}$$

**EXTRA!**
Digital roots can be useful for any number that is divided by 9. For example:

$$24 \div 9 = 2, \text{ remainder } 6$$

The digital root of 24 is 2 + 4 = **6**. Six is the remainder when 24 is divided by 9.

$$19 \div 9 = 2, \text{ remainder } 1$$

The digital root of 19 is 1 + 9 = 10, 1 + 0 = **1**. One is the remainder when 19 is divided by 9.

Investigate if this works for all numbers. What about numbers in the 9 times table?

# Checking calculations

Digital roots can be used to check quickly that the answers to addition sums are correct.

For example:

| | | |
|---|---|---|
| 4 5 1 | Digital root of 451 is **1** | 1 |
| + 3 6 8 | Digital root of 368 is **8** | +   8 |
| 8 1 9 | Digital root of 819 is **9** | 9 |

If the answer is correct its digital root will equal the total of the digital roots of the two numbers being added. Try some other numbers here.

Find out what happens when you add three or four numbers together. Does it still work? Try it here and see.

---

**EXTRA!**
How can digital roots help you to check that
subtraction and multiplication sums are correct?
Try a few and see.

---

How to be Brilliant at Algebra

# Exploring patterns, 1

Look closely at the patterns these sums make. Use a calculator to help you continue the patterns further. Record your answers on a separate sheet of paper.

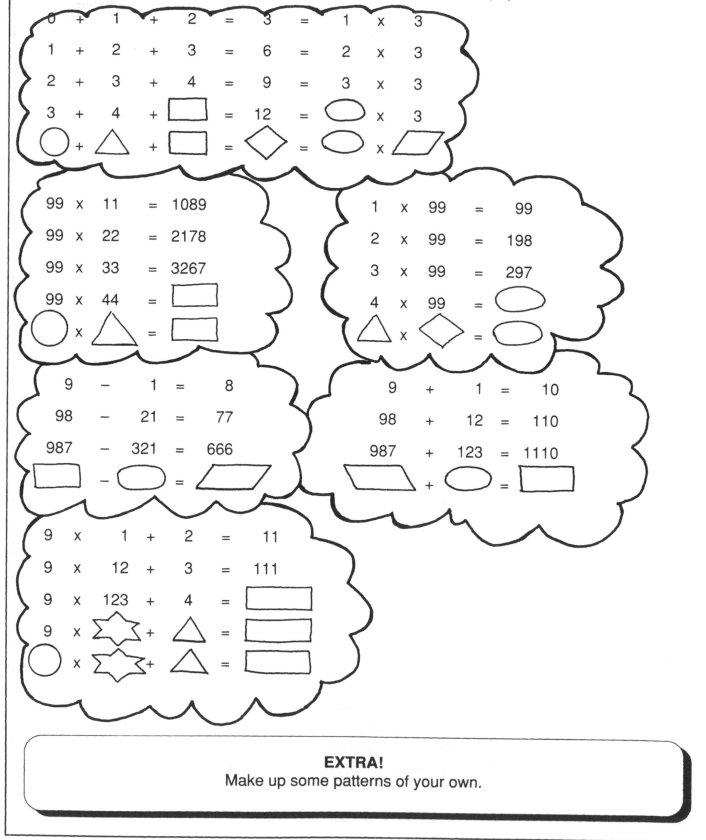

---

# Exploring patterns, 2

Investigate this number pattern. Use a calculator to help you.

| | | | | | | | |
|---|---|---|---|---|---|---|---|
| 2 x 2 | = | 1 x 1 | + | 2 x 1 | + | 1 |
| 3 x 3 | = | 2 x 2 | + | 2 x 2 | + | 1 |
| 4 x 4 | = | 3 x 3 | + | 2 x 3 | + | 1 |
| 5 x 5 | = | 4 x 4 | + | 2 x 4 | + | 1 |
| 6 x 6 | = | 5 x 5 | + | 2 x 5 | + | 1 |
| $n$ x $n$ | = | $(n-1)$ x $(n-1)$ | + | 2 x $(n-1)$ | + | 1 |

so

| | | | | | | |
|---|---|---|---|---|---|---|
| 20 x 20 | = | 19 x 19 | + | 2 x 19 | + | 1 |

Continue these patterns further then describe them using algebra.

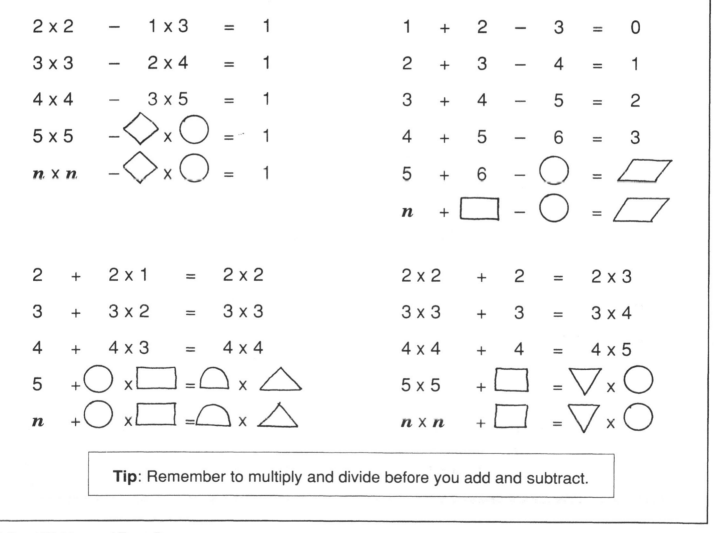

Tip: Remember to multiply and divide before you add and subtract.

How to be Brilliant at Algebra

# Growing shapes

Make a triangle using 3 toothpicks.

Add another triangle using 2 more toothpicks.

How many toothpicks do you need to make 3, 4, 5, 6, 7 and 8 triangles?
Complete the table.

| Number of triangles | 1 | 2 | 3 | 4 | 5 | 6 | 7 | 8 |
|---|---|---|---|---|---|---|---|---|
| Number of toothpicks | | | | | | | | |

Predict how many toothpicks you would need for 9 triangles. Can you make up a rule to say how many toothpicks will be needed for any number of triangles?

> **Tip**: It may help to represent the number of triangles by '*t*'.

Make a square using 4 toothpicks.

Add another square by using 3 more toothpicks.

How many toothpicks do you need to make 3, 4, 5, 6, 7 and 8 squares?
Complete this table.

| Number of squares | 1 | 2 | 3 | 4 | 5 | 6 | 7 | 8 |
|---|---|---|---|---|---|---|---|---|
| Number of toothpicks | | | | | | | | |

---

### EXTRA!
Predict how many sticks you would need for 9 squares.
Can you find a rule to say how many sticks will be
needed for any number of squares?

---

# Equations, 1

An equation is a statement using an equals sign.
For example:

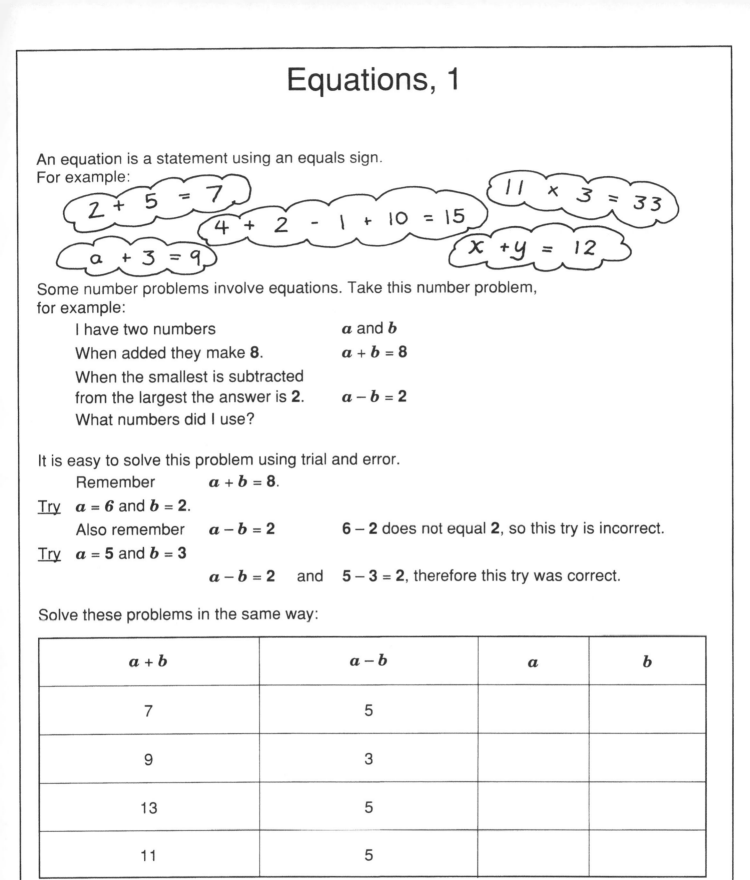

$$2 + 5 = 7$$
$$4 + 2 - 1 + 10 = 15$$
$$11 \times 3 = 33$$
$$a + 3 = 9$$
$$x + y = 12$$

Some number problems involve equations. Take this number problem, for example:

| I have two numbers | $a$ and $b$ |
| --- | --- |
| When added they make **8**. | $a + b = 8$ |
| When the smallest is subtracted from the largest the answer is **2**. | $a - b = 2$ |
| What numbers did I use? | |

It is easy to solve this problem using trial and error.

Remember $a + b = 8$.

<u>Try</u> $a = 6$ and $b = 2$.

Also remember $a - b = 2$       **6 – 2** does not equal **2**, so this try is incorrect.

<u>Try</u> $a = 5$ and $b = 3$

$a - b = 2$   and   **5 – 3 = 2**, therefore this try was correct.

Solve these problems in the same way:

| $a + b$ | $a - b$ | $a$ | $b$ |
| --- | --- | --- | --- |
| 7 | 5 | | |
| 9 | 3 | | |
| 13 | 5 | | |
| 11 | 5 | | |

---

**EXTRA!**
Make up some more of your own and try them on a friend.

---

How to be Brilliant at Algebra

# Equations, 2

Use trial and error to solve this problem:

| | |
|---|---|
| I have two numbers: | $a$ and $b$ |
| When multiplied they make **27**: | $a \times b = 27$ |
| When the largest is divided by the smallest the answer is **3**: | $a \div b = 3$ |
| What numbers did I use? | |

My solution:

Remember $a \times b = 27$

Try $a = 9$ and $b = 3$

Also remember $a \div b = 3$ $9 \div 3 = 27$

so $a = 9$ and $b = 3$.

*This try is correct!*

Solve these problems:

| $a \times b$ | $a \div b$ | $a$ | $b$ |
|:---:|:---:|:---:|:---:|
| 24 | 6 | | |
| 32 | 8 | | |
| 63 | 7 | | |
| 48 | 3 | | |
| 45 | 5 | | |

**EXTRA!**
Make up some more of your own and try them out on a friend.

How to be Brilliant at Algebra

# Pyramid numbers

Investigate how this pyramid of numbers was created:

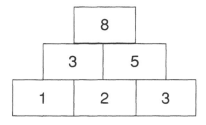

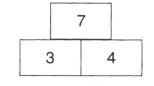
Rearrange the bottom row and calculate the new top number. Rearrange again and investigate what are the largest and smallest top numbers you can make?

How many different pyramids can you make using the same three bottom numbers?

Do the same for a pyramid with four numbers in the bottom row. Try using only odd or only even numbers in the bottom row.

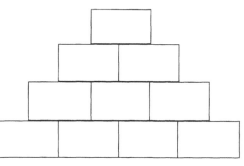

What are the largest and smallest top numbers you can make? What position do the highest numbers on the bottom row have when you make the largest top numbers?

**EXTRA!**
Try to find the equations to show why this happens.

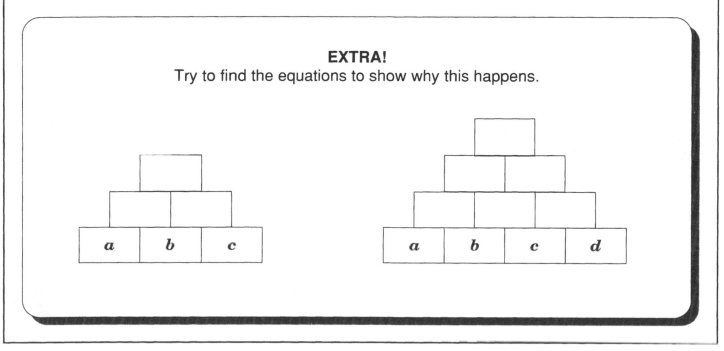

# What are the numbers?

Work with a friend. You will need two dice.

Secretly throw the dice to get two numbers and write them down. Label one number '$a$' and the other '$b$'. Follow the rule below and tell your friend what the rule is and what the answer is.

| Rule | Equation |
|------|----------|
| Add the two numbers together and add 3 | $a + b + 3$ |

Your friend should record the equation and the answer. For example: $a + b + 3 = 10$

He should try to work out what numbers you threw with the dice.

Try the same with these rules.

| Rule | Equation | Record your equations and answers here |
|------|----------|----------------------------------------|
| Multiply the two numbers together and take away 5 | $a \times b - 5$ | |
| Take away the first number from the second number and add 4 | $b - a + 4$ | |
| Double the first number and add the second number | $2 \times a + b$ | |
| Take 1 away from the second number and add the first number | $(b - 1) + a$ | |

**EXTRA!**
Try some rules of your own.

# Equations, 3

If    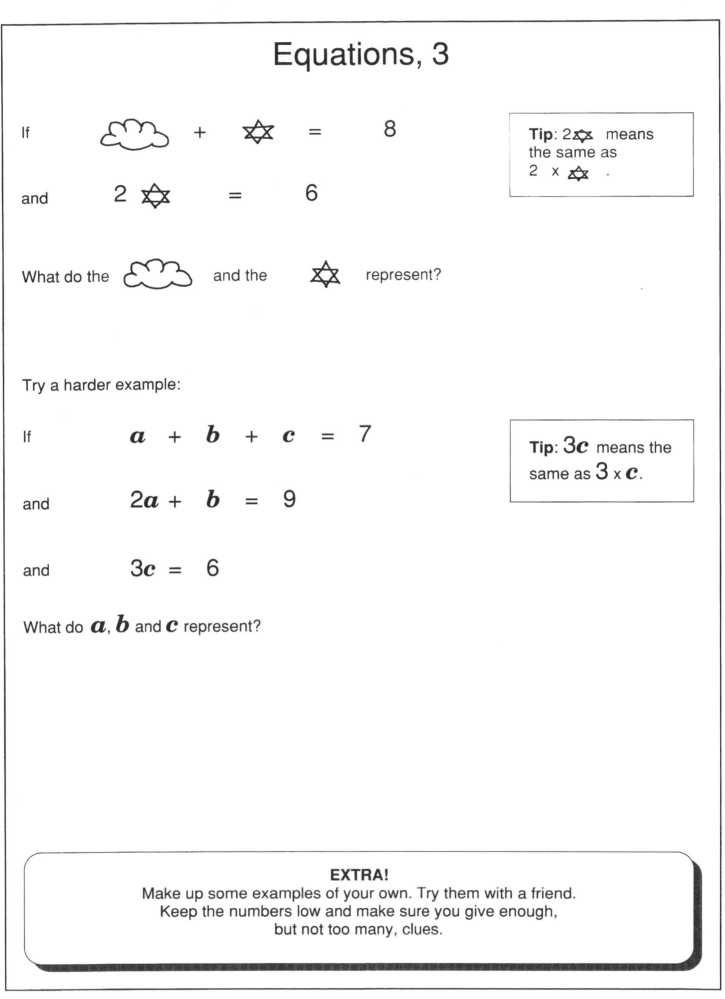 + ⬡ = 8

and    2 ⬡ = 6

**Tip:** 2⬡ means the same as 2 x ⬡ .

What do the ☁ and the ⬡ represent?

Try a harder example:

If    $a + b + c = 7$

and    $2a + b = 9$

and    $3c = 6$

**Tip:** $3c$ means the same as $3 \times c$.

What do $a$, $b$ and $c$ represent?

**EXTRA!**
Make up some examples of your own. Try them with a friend.
Keep the numbers low and make sure you give enough,
but not too many, clues.

How to be Brilliant at Algebra

# 121

Is 121 a square number? You can arrange 121 dots in a square pattern.

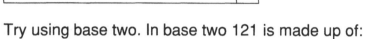

121 is made up of one hundred, two tens and one unit in base ten.

| 10 x 10 = **100** | 1 x 10 = **10** |
| --- | --- |
| 1 x 10 = **10** | 1 |

**100 + 10 + 10 + 1 = 121**

Try using base two. In base two 121 is made up of:

| 2 x 2 = **4** | 1 x 2 = **2** |
| --- | --- |
| 1 x 2 = **2** | 1 |

**4 + 2 + 2 + 1 = 9**

9 is a square number in base ten.

Investigate 121 in other bases. Is 121 in other bases always a square number in base ten?

# Digital root patterns

**Tip**: Digital roots are calculated by adding the individual digits of a number. For example:

|  | | | |
|---|---|---|---|
| | **132** | digital root is | 1 + 3 + 2 = **6** |
| and | **276** | digital root is | 2 + 7 + 6 = 15 |
| | | | 1 + 5 = **6** |

Find the digital roots of the numbers in all the times tables up to 9. Write them in the table below.

| x | 1 | 2 | 3 | 4 | 5 | 6 | 7 | 8 | 9 | 10 |
|---|---|---|---|---|---|---|---|---|---|---|
| 1 | | | | | | | | | | |
| 2 | | | | | | | | | | |
| 3 | | | | | | | | | | |
| 4 | | | | | | | | | | |
| 5 | | | | | | | | | | |
| 6 | | | | | | | | | | |
| 7 | | | | | | | | | | |
| 8 | | | | | | | | | | |
| 9 | | | | | | | | | | |

*Look for patterns. What do you notice?*

## EXTRA!

Investigate the spirals you can draw using the digital root sequences. Choose a starting position on a large sheet of squared paper. Draw a line the same number of squares long as the first number in the sequence. The next line should be drawn having turned the paper clockwise through 90 degrees and be the length of the second number in the sequence. Continue until you return to your starting point. Does this always work?

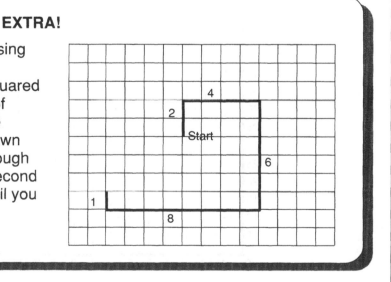

# Constant perimeter

**Tip:**
The distance round the outside of a shape is called the **perimeter**.

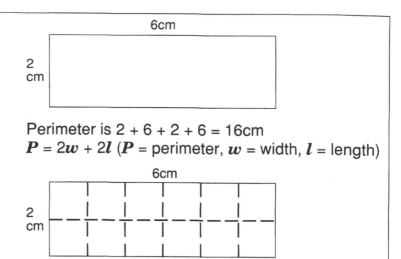

Perimeter is 2 + 6 + 2 + 6 = 16cm
$P = 2w + 2l$ ($P$ = perimeter, $w$ = width, $l$ = length)

The space in the middle of the rectangle is called the **area**.

Area is 2 x 6 = 12cm$^2$
$A = w \times l$

Investigate the rectangles you can make that all have a perimeter of 36cm. Use only whole centimetres. Record your results in this table.

| width | 1 | 2 | 3 | 4 | 5 | 6 | 7 | 8 | 9 | 10 | 11 | 12 | 13 | 14 | 15 | 16 | 17 |
|---|---|---|---|---|---|---|---|---|---|---|---|---|---|---|---|---|---|
| length | | | | | | | | | | | | | | | | | |
| area | | | | | | | | | | | | | | | | | |

Plot the width and the area on the graph below. What do you notice? Try again using a different perimeter.

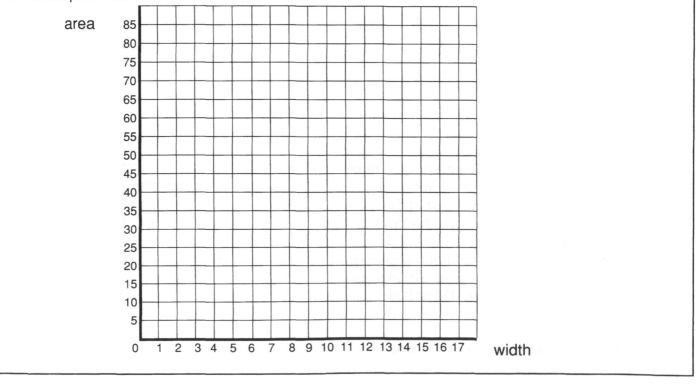

# Balancing

Each of these groups balances each other. Use this information to solve the problems.

**Tip**: Look for the same groups of shapes.

How to be Brilliant at Algebra

# Straight line graphs

Number patterns can be drawn on graphs. It helps to make a table first.
The table for 'double the number' would be:

| $x$ | 1 | 2 | 3 | 4 | 5 | 6 | 7 | 8 | 9 | 10 |
|-----|---|---|---|---|---|---|---|---|---|----|
| $2x$ | 2 | 4 | 6 | 8 | 10 | 12 | 14 | 16 | 18 | 20 |

The pattern has been plotted on the graph below.

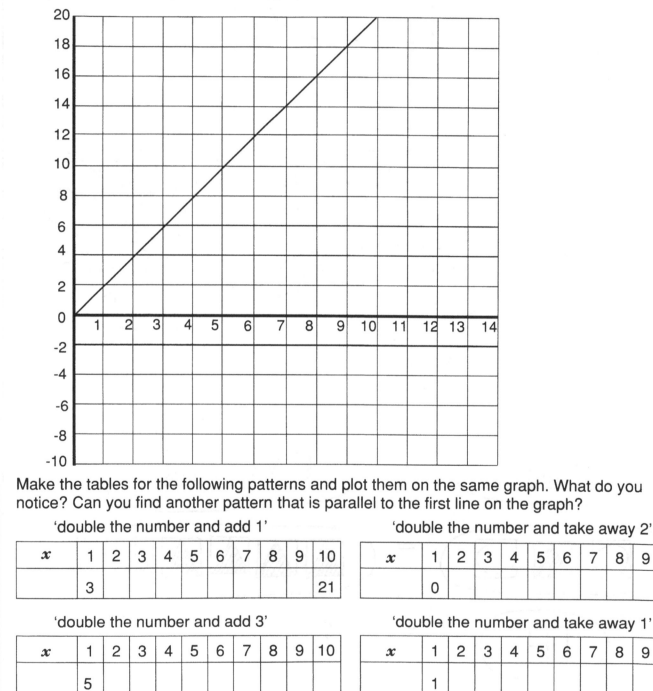

Make the tables for the following patterns and plot them on the same graph. What do you notice? Can you find another pattern that is parallel to the first line on the graph?

'double the number and add 1'

| $x$ | 1 | 2 | 3 | 4 | 5 | 6 | 7 | 8 | 9 | 10 |
|-----|---|---|---|---|---|---|---|---|---|----|
|  | 3 |  |  |  |  |  |  |  |  | 21 |

'double the number and take away 2'

| $x$ | 1 | 2 | 3 | 4 | 5 | 6 | 7 | 8 | 9 | 10 |
|-----|---|---|---|---|---|---|---|---|---|----|
|  | 0 |  |  |  |  |  |  |  |  |  |

'double the number and add 3'

| $x$ | 1 | 2 | 3 | 4 | 5 | 6 | 7 | 8 | 9 | 10 |
|-----|---|---|---|---|---|---|---|---|---|----|
|  | 5 |  |  |  |  |  |  |  |  |  |

'double the number and take away 1'

| $x$ | 1 | 2 | 3 | 4 | 5 | 6 | 7 | 8 | 9 | 10 |
|-----|---|---|---|---|---|---|---|---|---|----|
|  | 1 |  |  |  |  |  |  |  |  |  |

# Curved line graphs

Some number patterns make curved lines when drawn on a graph.
The example below shows the graph of 'square the number'.

| $x$ | 1 | 2 | 3 | 4 | 5 | 6 | 7 | 8 |
|-----|---|---|---|----|----|----|----|----|
| $x^2$ | 1 | 4 | 9 | 16 | 25 | 36 | 49 | 64 |

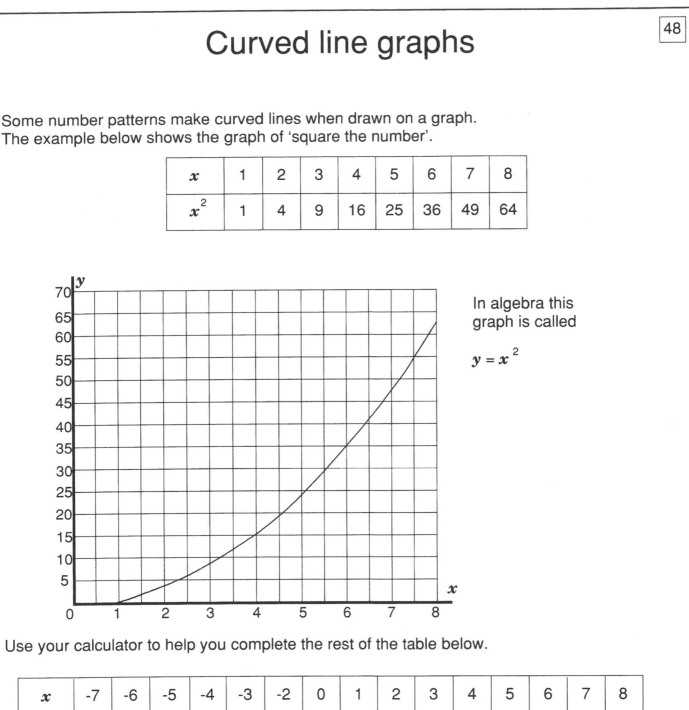

In algebra this graph is called

$$y = x^2$$

Use your calculator to help you complete the rest of the table below.

| $x$ | -7 | -6 | -5 | -4 | -3 | -2 | 0 | 1 | 2 | 3 | 4 | 5 | 6 | 7 | 8 |
|-----|----|----|----|----|----|----|---|---|---|---|----|----|----|----|----|
| $x^2$ | | | | | | | | 1 | 4 | 9 | 16 | 25 | 36 | 49 | 64 |

What do you notice happens when you multiply two negative numbers together? Plot the number pattern on the graph paper resource sheet (page 48). What do you notice about the right and left sides of the graph?

Make tables for the following patterns and plot them on the same graph. What do you notice? Can you find another pattern that is parallel to the first line on the graph?

'square the number and add 5 '          'square the number and subtract 5'

'square the number and add 10'          'square the number and subtract 8'

How to be Brilliant at Algebra

# Prime numbers

All numbers are either **composite** or **prime**. **Composite** numbers can be arranged in a rectangle with more than one row. For example:

12

**12 is a composite number.**

**Prime** numbers can only be arranged in single rows. A prime number is said to be divisible only by one and itself. For example:

7

**7 is a prime number.**

| 1 | 2 | 3 | 4 | 5 | 6 |
|---|---|---|---|---|---|
| 7 | 8 | 9 | 10 | 11 | 12 |
| 13 | 14 | 15 | 16 | 17 | 18 |
| 19 | 20 | 21 | 22 | 23 | 24 |
| 25 | 26 | 27 | 28 | 29 | 30 |
| 31 | 32 | 33 | 34 | 35 | 36 |
| 37 | 38 | 39 | 40 | 41 | 42 |
| 43 | 44 | 45 | 46 | 47 | 48 |
| 49 | 50 | 51 | 52 | 53 | 54 |
| 55 | 56 | 57 | 58 | 59 | 60 |
| 61 | 62 | 63 | 64 | 65 | 66 |
| 67 | 68 | 69 | 70 | 71 | 72 |
| 73 | 74 | 75 | 76 | 77 | 78 |
| 79 | 80 | 81 | 82 | 83 | 84 |
| 85 | 86 | 87 | 88 | 89 | 90 |
| 91 | 92 | 93 | 94 | 95 | 96 |
| 97 | 98 | 99 | 100 | 101 | 102 |

Two is a prime number, but all the other even numbers are not. Circle the number 2 and cross out all the other even numbers on the chart.

The multiples of 3 are in two columns. Circle the number 3 and cross out all the others.

The multiples of 5 are in diagonals. Circle the number 5. Cross out the other multiples of 5.

Keep going through all the times tables until only the prime numbers are left.

What do you notice?

## EXTRA!

There is no formula for calculating all the prime numbers. Try:

$$n^2 - n + 11$$

| $n$ | 1 | 2 | 3 | 4 | 5 | 6 | 7 | 8 | 9 | 10 | 11 |
|---|---|---|---|---|---|---|---|---|---|---|---|
| $n^2$ | 1 | 4 | 9 | | | | | | | | |
| $n^2 - n$ | 0 | 2 | 6 | | | | | | | | |
| $n^2 - n + 11$ | 11 | 13 | 17 | | | | | | | | |

It will not work for $n = 11$. Why not?

# Fibonacci numbers, 1

Fibonacci's number pattern is formed by adding the two previous numbers together to get the next term. Continue Fibonacci's pattern.

| 1 | 1 | 2 | 3 | 5 | 8 | | | | | | | 233 |

1st term  2nd term  3rd term

■ Calculate the sum of the first three terms. Compare your total to the 5th term. Now compare the sum of the first four terms with the 6th term. Does this always work?

■ Look for patterns of even and odd numbers.

■ Look for multiples of 3, 4 and 5. Is there a pattern?

■ Choose some groups of three terms. Square the middle term and multiply the two outside terms together. What do you notice? Does it always work?

$$2 \quad 3 \quad 5$$

$$3 \times 3 = 9$$

$$2 \times 5 = 10$$

■ Choose some groups of four terms. Multiply the inside terms together and the outside numbers together. Compare them. What do you notice? Try for other groups of four terms.

$$5 \quad 8 \quad 13 \quad 21$$

$$8 \times 13 = 104$$

$$5 \times 21 = 105$$

# Fibonacci numbers, 2

**Tip**: The Fibonacci sequence goes:

| 1 | 1 | 2 | 3 | 5 | 8 | 13 | 21 | 34 | 55 | 89 | 144 | 233 |

To find a digital root you total the individual digits of a number. Keep going until you reach a single digit.

For example: $89 \rightsquigarrow 8 + 9 = 17 \rightsquigarrow 1 + 7 = 8$  8 the digital root of 89.

Use a calculator to find the first 24 terms of the Fibanacci sequence.

| 1 | 1 | 2 | 3 |
|---|---|---|---|
| 5 | | | |
| | | | |
| | | | |
| | | | |
| | | | 46368 |

Calculate the digital roots of the first 24 terms. For example:

$$4\ 6\ 3\ 6\ 8 \qquad 4 + 6 + 3 + 6 + 8 = 27$$
$$2 + 7 \qquad\quad = \quad 9 \qquad \text{digital root}$$

Investigate the patterns in the other digital roots in the sequence.

---

**EXTRA!**
Starting with the second 1 and counting on six terms, what pattern can you see?

---

# Groups of five

Fibonacci was a famous Italian mathematician. He lived between 1170 and 1250. He wrote books on algebra. He particularly liked one number pattern. It is formed by adding the two previous numbers together to get the next term.

Investigate groups of five numbers created by using Fibonacci's rule.

1 Choose two starting numbers.

| 3 | 4 | | | |
|---|---|---|---|---|

2 Add them together to make the third term.

| 3 | 4 | 7 | | |
|---|---|---|---|---|

3 Complete the group in the same way.

| 3 | 4 | 7 | 11 | 18 |
|---|---|---|---|---|

Make several groups of five in this way. Give your friend only the first and last number and see if she can work out what the missing numbers are.

It is easy if you work it out algebraically. What are the last two terms in this group?

| $a$ | $b$ | $a+b$ | | |
|---|---|---|---|---|

Use this information to help you solve these groups.

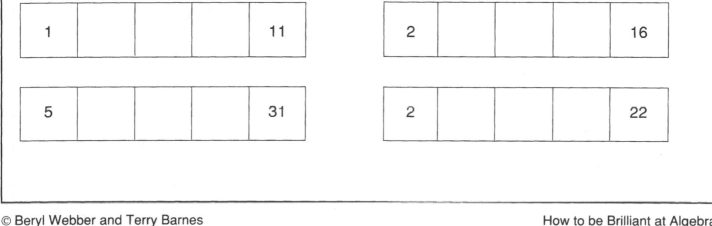

| 1 | | | | 11 |
|---|---|---|---|---|

| 2 | | | | 16 |
|---|---|---|---|---|

| 5 | | | | 31 |
|---|---|---|---|---|

| 2 | | | | 22 |
|---|---|---|---|---|

# Pascal's triangle

This number triangle was known to the Chinese in the fourteenth century. Pascal was a French philosopher who was born in 1623 and died in 1662. He made the triangle famous in Europe.

Investigate the triangle and see how it is formed. How many number patterns can you find?

**EXTRA!**
Try colouring all the even numbers one colour and the odd numbers another.
Look at the pattern formed.

Pascal's triangle

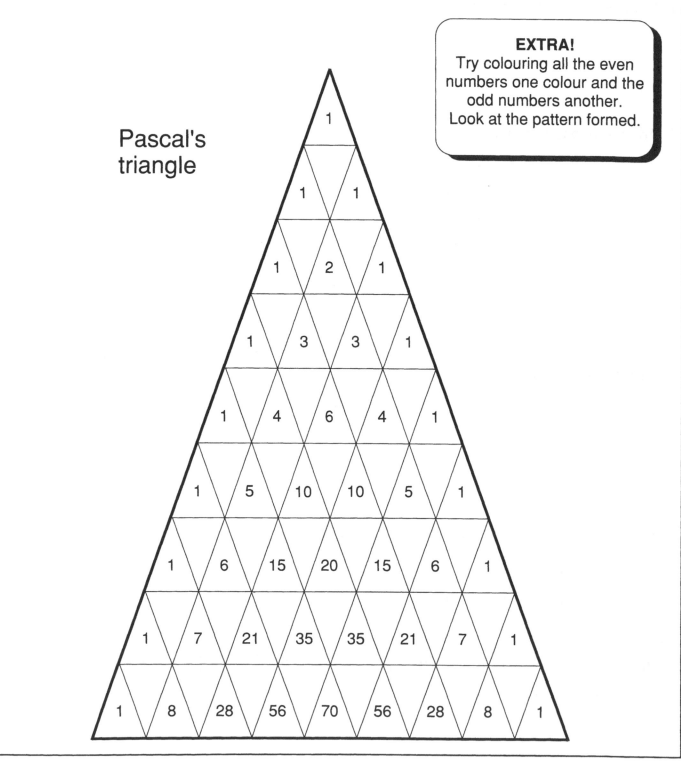

# The golden ratio

The golden ratio is very pleasing to the eye. It has been used for centuries for the proportions of buildings. It is a number that goes on forever. It starts like this: 1.618. A rectangle that is in the golden ratio would have a length that was 1.618 times the width. A4 paper is very nearly in the golden ratio. Measure it and see.

This is how to construct a golden rectangle:

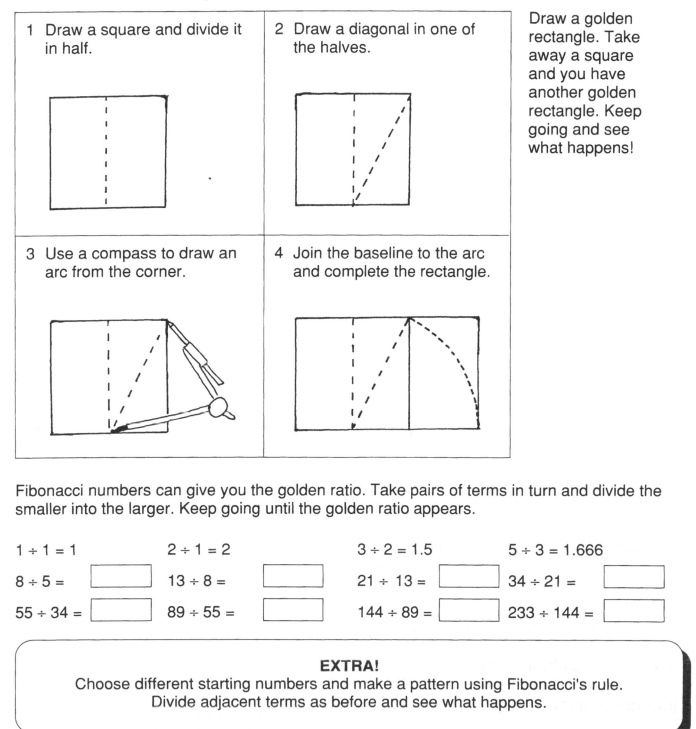

1 Draw a square and divide it in half.

2 Draw a diagonal in one of the halves.

3 Use a compass to draw an arc from the corner.

4 Join the baseline to the arc and complete the rectangle.

Draw a golden rectangle. Take away a square and you have another golden rectangle. Keep going and see what happens!

Fibonacci numbers can give you the golden ratio. Take pairs of terms in turn and divide the smaller into the larger. Keep going until the golden ratio appears.

$1 \div 1 = 1$          $2 \div 1 = 2$          $3 \div 2 = 1.5$          $5 \div 3 = 1.666$

$8 \div 5 =$ ☐          $13 \div 8 =$ ☐          $21 \div 13 =$ ☐          $34 \div 21 =$ ☐

$55 \div 34 =$ ☐          $89 \div 55 =$ ☐          $144 \div 89 =$ ☐          $233 \div 144 =$ ☐

---

**EXTRA!**
Choose different starting numbers and make a pattern using Fibonacci's rule.
Divide adjacent terms as before and see what happens.

How to be Brilliant at Algebra

Name _____

# I can ...

| | Date |
|---|---|
| identify patterns | |
| recognize the effects of inverse operations | |
| identify odd and even numbers | |
| predict number patterns | |
| work systematically | |
| make logical deductions | |
| identify prime numbers | |
| investigate square numbers | |
| use brackets in calculations | |
| investigate digital roots | |
| investigate number patterns | |
| generalize number patterns | |
| use simple word formulae | |
| use simple symbolic formulae | |
| solve simple equations | |
| use algebra to solve problems | |
| use co-ordinates | |
| plot number patterns on a graph | |
| interpret straight line graphs | |
| interpret curved line graphs | |

# Calendar resource sheet

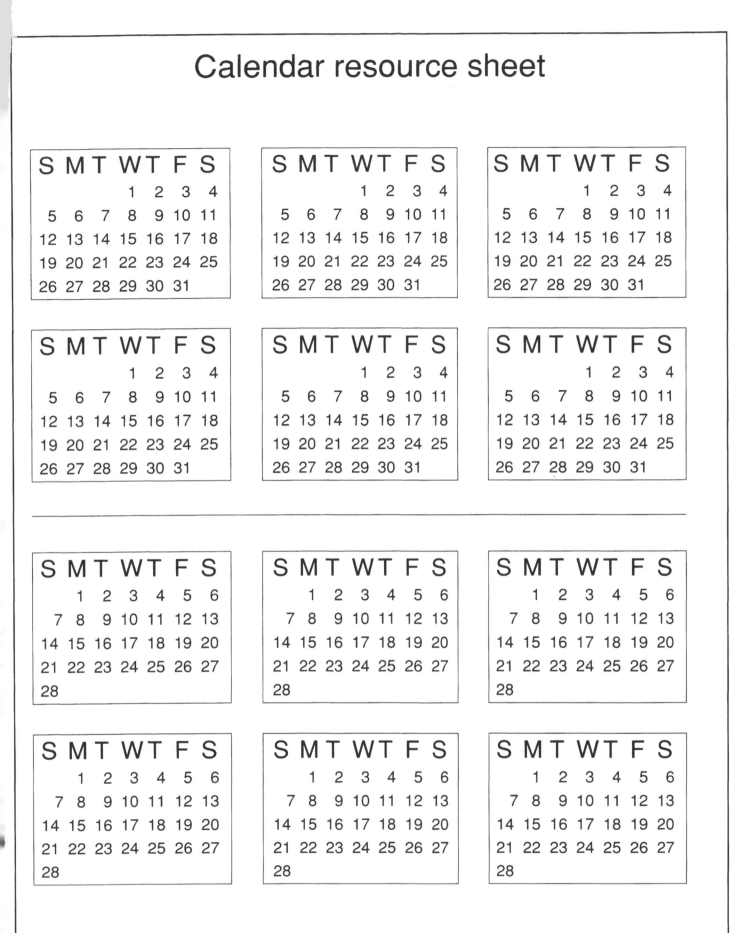

| S | M | T | W | T | F | S |
|---|---|---|---|---|---|---|
| | | | 1 | 2 | 3 | 4 |
| 5 | 6 | 7 | 8 | 9 | 10 | 11 |
| 12 | 13 | 14 | 15 | 16 | 17 | 18 |
| 19 | 20 | 21 | 22 | 23 | 24 | 25 |
| 26 | 27 | 28 | 29 | 30 | 31 | |

| S | M | T | W | T | F | S |
|---|---|---|---|---|---|---|
| | | | 1 | 2 | 3 | 4 |
| 5 | 6 | 7 | 8 | 9 | 10 | 11 |
| 12 | 13 | 14 | 15 | 16 | 17 | 18 |
| 19 | 20 | 21 | 22 | 23 | 24 | 25 |
| 26 | 27 | 28 | 29 | 30 | 31 | |

| S | M | T | W | T | F | S |
|---|---|---|---|---|---|---|
| | | | 1 | 2 | 3 | 4 |
| 5 | 6 | 7 | 8 | 9 | 10 | 11 |
| 12 | 13 | 14 | 15 | 16 | 17 | 18 |
| 19 | 20 | 21 | 22 | 23 | 24 | 25 |
| 26 | 27 | 28 | 29 | 30 | 31 | |

| S | M | T | W | T | F | S |
|---|---|---|---|---|---|---|
| | | | 1 | 2 | 3 | 4 |
| 5 | 6 | 7 | 8 | 9 | 10 | 11 |
| 12 | 13 | 14 | 15 | 16 | 17 | 18 |
| 19 | 20 | 21 | 22 | 23 | 24 | 25 |
| 26 | 27 | 28 | 29 | 30 | 31 | |

| S | M | T | W | T | F | S |
|---|---|---|---|---|---|---|
| | | | 1 | 2 | 3 | 4 |
| 5 | 6 | 7 | 8 | 9 | 10 | 11 |
| 12 | 13 | 14 | 15 | 16 | 17 | 18 |
| 19 | 20 | 21 | 22 | 23 | 24 | 25 |
| 26 | 27 | 28 | 29 | 30 | 31 | |

| S | M | T | W | T | F | S |
|---|---|---|---|---|---|---|
| | | | 1 | 2 | 3 | 4 |
| 5 | 6 | 7 | 8 | 9 | 10 | 11 |
| 12 | 13 | 14 | 15 | 16 | 17 | 18 |
| 19 | 20 | 21 | 22 | 23 | 24 | 25 |
| 26 | 27 | 28 | 29 | 30 | 31 | |

| S | M | T | W | T | F | S |
|---|---|---|---|---|---|---|
| | 1 | 2 | 3 | 4 | 5 | 6 |
| 7 | 8 | 9 | 10 | 11 | 12 | 13 |
| 14 | 15 | 16 | 17 | 18 | 19 | 20 |
| 21 | 22 | 23 | 24 | 25 | 26 | 27 |
| 28 | | | | | | |

| S | M | T | W | T | F | S |
|---|---|---|---|---|---|---|
| | 1 | 2 | 3 | 4 | 5 | 6 |
| 7 | 8 | 9 | 10 | 11 | 12 | 13 |
| 14 | 15 | 16 | 17 | 18 | 19 | 20 |
| 21 | 22 | 23 | 24 | 25 | 26 | 27 |
| 28 | | | | | | |

| S | M | T | W | T | F | S |
|---|---|---|---|---|---|---|
| | 1 | 2 | 3 | 4 | 5 | 6 |
| 7 | 8 | 9 | 10 | 11 | 12 | 13 |
| 14 | 15 | 16 | 17 | 18 | 19 | 20 |
| 21 | 22 | 23 | 24 | 25 | 26 | 27 |
| 28 | | | | | | |

| S | M | T | W | T | F | S |
|---|---|---|---|---|---|---|
| | 1 | 2 | 3 | 4 | 5 | 6 |
| 7 | 8 | 9 | 10 | 11 | 12 | 13 |
| 14 | 15 | 16 | 17 | 18 | 19 | 20 |
| 21 | 22 | 23 | 24 | 25 | 26 | 27 |
| 28 | | | | | | |

| S | M | T | W | T | F | S |
|---|---|---|---|---|---|---|
| | 1 | 2 | 3 | 4 | 5 | 6 |
| 7 | 8 | 9 | 10 | 11 | 12 | 13 |
| 14 | 15 | 16 | 17 | 18 | 19 | 20 |
| 21 | 22 | 23 | 24 | 25 | 26 | 27 |
| 28 | | | | | | |

| S | M | T | W | T | F | S |
|---|---|---|---|---|---|---|
| | 1 | 2 | 3 | 4 | 5 | 6 |
| 7 | 8 | 9 | 10 | 11 | 12 | 13 |
| 14 | 15 | 16 | 17 | 18 | 19 | 20 |
| 21 | 22 | 23 | 24 | 25 | 26 | 27 |
| 28 | | | | | | |

How to be Brilliant at Algebra

9 781897 675052